You Wouldn't Want to...

BE IN A TANK IN

WWII!

Written by Roger Canavan
Illustrated by David Antram

Hatch

Contents

Introduction	5	D-Day and beyond	24
A long history	6	New battlefields	26
The Iron Triangle	8	Look to the future	28
Tested in battle	10	More tank topics	30
Another war looming?	12	Life in a tank	32
Build! Build! Build!	14	Tank timeline	34
Ready to roll	16	Map of WW II tank battles	36
Taking on the Desert Fox	18	Glossary	37
Clash of armour	20	Index	39
Funny business?	22		

Introduction

There's no doubt about it – being in a tank crew during the Second World War puts you at the cutting edge of action. As soon as you begin training, you soon realise that this powerful, armoured weapon is able to intimidate enemies, win battles, and bulldoze its way to victory.

It's taken hundreds of years, and plenty of false starts, for tanks to reach their present-day size, shape and power. Designs have changed along the way, keeping pace with the evolving nature of wars and weapons. But successful tanks share three key advantages: their armour provides protection, their tracks provide mobility, and their explosive weapons provide enough firepower to defeat enemy forces.

The tank crew endure difficult conditions: cramped spaces, extreme temperatures, and the risk of an even bigger tank waiting for them over the next hill.

No, you definitely wouldn't want to be in a tank in the Second World War, but if you were, you'd be grateful for the company. After all, a tank is only as effective as the crew that operates it.

Wait a minute – the map says we're on the wrong hill!

And the front line is in the valley beyond this range of hills.

A long history

Are you looking to gain an edge in battle, or to discover new methods of protection? You wouldn't be alone. For centuries, military designers have come up with ideas to do just that. Some were brilliant; others were far-fetched and unworkable. Leonardo da Vinci seemed to be both! But each one brought us another step closer to the tank as a lethal combination of mobility, protection and firepower. Engineers and inventors learned from these earlier 'false starts', as they developed ideas that would eventually evolve into the tanks we recognise today.

This contraption could never be built!

DA VINCI'S DESIGN
In 1487, Leonardo Da Vinci designed a fighting machine that would 'enter the closed ranks of the enemy... so that our infantry will be able to follow quite unharmed'. But the idea didn't catch on.

BATTLE CARS
Designed in 1855, the James Cowen battle car combined a steam engine, six cannons and retractable scythes. The British Army wasn't interested, and Cowen complained that its leaders 'didn't understand engineering'.

"Wrong – I've made a model right here."

Handy Hint
A good tank design features angled sides. This provides better protection without the need for more metal.

IRONCLADS
In 1862, two armour-plated ships, known as 'Ironclads', fight a famous battle in the American Civil War. Could land versions be far behind?

LITERARY INSPIRATIONS
A 1903 story by British author H. G. Wells imagined 'land ironclads' overcoming trench defences. It was probably inspired by the 'Ironclad' naval battle but also predicted some of the trench warfare to come.

The Iron Triangle

It was becoming clear that armour, which once protected knights, could be the key to modern battlefield combat. It would put the 'P' in the Iron Triangle of mobility, protection and firepower that inspired design after design. The British agreed to test one of the new tank designs early on in the First World War. 'Little Willie' was a bit clunky and tended to get stuck in trenches. After this disappointing test run, people wondered whether there could be a future for tanks... or had the idea already got stuck in the mud?

TANK DESIGNS SCRAPPED
On the eve of the First World War, several tank designs covered the three points of the Iron Triangle... but perhaps they were too ahead of their time.

THE HORNSBY TRACTOR
The Tank Museum's Hornsby Tractor, from 1910, is the only one left in the world. Its design highlighted the 'M' (mobility) of the Iron Triangle. Real tanks were just a few years away.

In 1911, Gunther Burstyn (an officer in the Austro-Hungarian Army) designed what looks like a modern tank, complete with a gun turret to bring 'F' (firepower). You guessed it: the idea was rejected by the German and Austrian armies.

Handy Hint

Remember that your new weapon design needs to convince the government – they'll foot the bill!

AUSTRALIAN INVENTIONS

Lancelot de Mole, an Australian engineer and inventor, submitted plans for an armoured tracked vehicle to the British War Office, but it was rejected in 1911... and again in 1914. Is there a pattern here?

TEST RUN

Little Willie's test run in 1915 wasn't a huge success. It chugged along at less than walking speed and struggled on slopes. But officers saw that with improvements it could be the basis for an effective fighting machine.

Hmm. This seems awfully slow and clunky.

But I think the idea has potential!

Tested in battle

You can recall the excitement when war was first declared in the late summer of 1914. But the cries of 'It'll all be over by Christmas' soon faded. The war chugged along, settling into a series of deadly battles in which neither side gained much of an advantage. Soldiers spend most of their time in deep, muddy trenches. It's grim on the battlefront, but the British Army has come up with a game-changer – and you're part of it!

It's time for tanks to make their first appearance in combat. You have been selected to join a crew inside one of the new fighting vehicles. Will tanks turn the tide, or will they become sitting ducks?

THE FIRST TANKS
Initially, the first tanks in battle unsettled troops on both sides. One British soldier recalled: 'We heard strange throbbing noises, and lumbering slowly towards us came three huge mechanical monsters such as we had never seen before.'

THE CAVALRY
During the First World War, the cavalry still played a large role, as they were the fastest part of the Army. Unlike horses, tanks didn't need to be fed or groomed. But they needed just as much – if not more – maintenance and care, not to mention fuel!

Yikes! Am I seeing things?

Another war looming?

You're a proud member of Britain's Royal Tank Corps, founded in the wake of the First World War. Army chiefs recognised the contribution of tanks to that victory, which is why the Corps was formed. Spirits were high back then, and some people even referred to the 'war to end all wars'.

Things in the 1930s seem different. You're enjoying your training with the tanks and other armoured equipment, but the daily news is less peaceful. Fighting has broken out in Spain's civil war, and there are military displays over in Germany. Adolf Hitler, the German leader, is telling his people that they must prepare for war. Your tank training begins to feel a lot more serious.

TANK BRAVERY
The motto on the Royal Tank Corps' cap badge – 'Fear Naught (nothing)' – referred to the might of the tanks as well as the team spirit of those who rode in them.

ARMS RACE
Germany was slow to adopt tanks in the First World War but by the mid-1930s was making up for things. Hitler and other leaders watched as dozens of new tanks paraded by at huge rallies.

TANK GROWTH
Conflicts involving tanks flared up across the globe in the 1930s. This Soviet T-26 tank helped the Republican forces in the Spanish Civil War. Germany, Italy and Japan were also sending tanks to other hot spots.

Handy Hint

If you're in the Royal Armoured Corps, take a close look at the German tanks you see on the movie screen – you might meet them before long.

A CHANGE IN MOOD
British tank crews trained in high spirits in the early 1930s – but before long the atmosphere would become deadly serious as another war became more likely.

IT'S WAR!
Teamwork underpinned Germany's dramatic early blitzkrieg ('lightning war') in northern Europe, defeating France in 1940. German tank crews were equipped with radios, which gave them the advantage in action and reaction as they swept across borders.

Build! Build! Build!

You've read how the German blitzkrieg sent the British Expeditionary Force retreating to the English Channel in May 1940. Many troops returned during the 'Miracle of Dunkirk', but most British vehicles and artillery were captured or destroyed. Only thirteen of 700 tanks returned to Britain.

Things are looking bad, but the country has turned to rebuilding the lost supplies. Your commanding officers in the Royal Tank Regiment have sent you on a tour of factories where different parts of tanks are produced. Lots of women have joined the workforce to help meet the growing demand. Some of the tanks roll out with messages of support (for the Allies) or taunts (for the enemy).

BUILDING A TANK
Some assembly lines concentrated on just one part of a tank; the parts would then be sent off to be assembled into tanks somewhere else.

MAVIS JONES
Mavis Jones was only 16 years old when she joined Newton Chambers, the tank manufacturer in Sheffield. She was assigned to the Drawing Office, where she produced many tank technical drawings and blueprints of Churchill tanks. These drawings had to be highly detailed and precise to be followed on the factory floor. Otherwise the finished tanks wouldn't work!

HELPED TO BUILD THIS TANK COME AND JOIN THEM?

Handy Hint
Your tank needs to be hoisted onto ships and to fit under railway bridges. Don't make it too heavy or too big!

"I hope this winds up under his Christmas tree!"

We Can Do It!

WOMEN AND WEAPONS
American women were also urged to join the war effort in producing tanks and other weapons after the US entered the war in 1941. 'Rosie the Riveter', with her upbeat slogan, was the symbol of the contribution.

FINISHING TOUCHES
Workers in tank factories enjoyed adding some extra touches before the finished tanks were sent off. There was no doubt about where this tank was heading.

Ready to roll

It's a tight squeeze for the crew of a typical tank. The combination of heavy armour, fuel tanks, weapon and driving controls, and ammunition leaves very little space to sit comfortably or lie back. Do you feel like standing up or stretching? Forget it – at least until the fighting stops and it's safe to get out.

Check out all the pieces that link together under the armour. Have you memorised where they all go, and what they connect to? Because just like those First World War crews, you'll have to be able to make repairs – sometimes hundreds of miles from the nearest mechanic. This image shows an American tank, the M4A4 Sherman, which was also used by other Allies throughout the war. Most tanks have similar features.

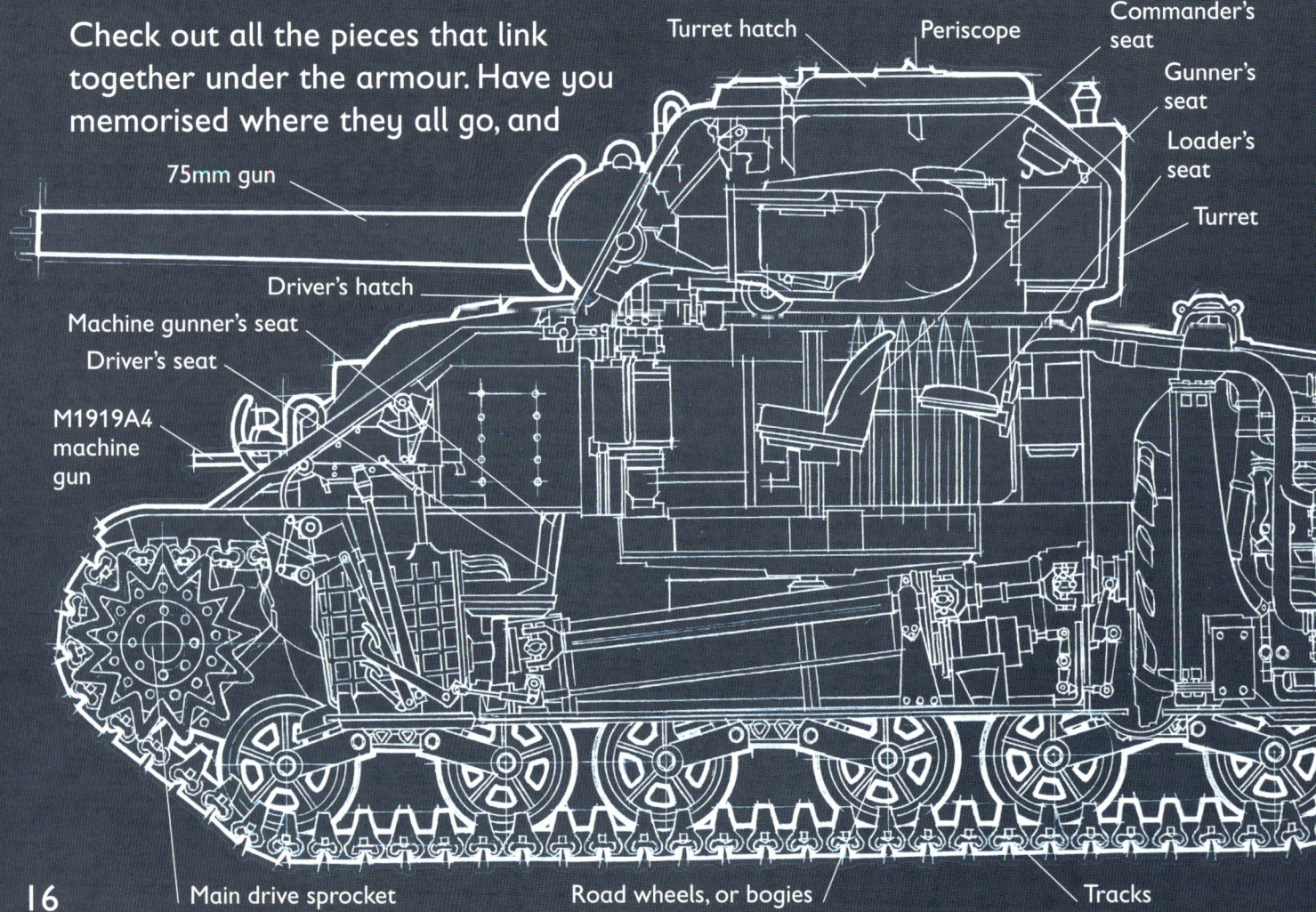

ROLES OF THE CREW

A typical tank crew had five members. Each had his special job, but they could take one another's place in an emergency. In charge was the commander, who issued orders. The driver had to steer the tank – often through difficult terrain and in bad weather with terrible visibility. The loader and gunner operated the main gun while the hull machine gunner's weapon was just beneath it.

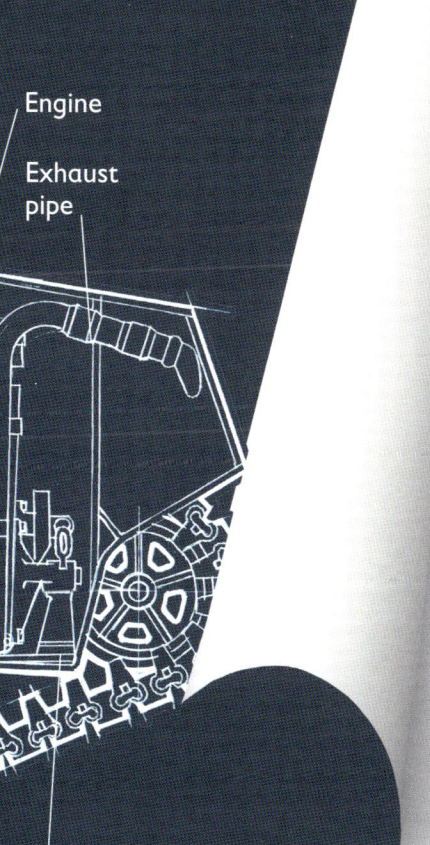

Engine

Exhaust pipe

Idler wheel

Commander

Handy Hint

Try to keep your muscles and joints flexible. It's a tight squeeze getting through the hatch – and you might need to escape in a hurry!

Loader Gunner Driver

Look. We're like one big, happy family.

Hull machine gunner

Taking on the Desert Fox

Your tank crew has been stationed in North Africa for just over a year, since 1941, and it feels like a long way from home. Scorching temperatures, scarce water, mosquitoes and sandstorms – a world away from London or Glasgow! It doesn't sound like Berlin or Munich either, but you're aware that a German armoured force – with a fearsome commander – is ready to do battle. Both sides aim to control that important desert stretch along the Mediterranean Sea, and with it the gateway to control supply routes to Asia with its oil reserves. Monty, your overall commander, decides to take on the tanks of the German Afrika Korps. It's the second battle of El Alamein, and there's a lot at stake.

IN CHARGE
The opposing commanders at El Alamein were viewed by their men as 'one of them' – brave leaders of a band of brothers.

BERNARD MONTGOMERY
Lieutenant-General Bernard Montgomery, or 'Monty' to his men, was always keen to be seen as one who got stuck in. He often boarded tanks to get a driver's eye view of the fighting.

ERWIN ROMMEL
General Erwin Rommel had been a key figure in Germany's defeat of France. By 1942 the 'Desert Fox' and his highly trained Afrika Korps had clashed several times with the British – including an earlier battle at El Alamein.

"Phew! Looks like things are heating up over there..."

Handy Hint
Desert fighting can take your tank across long distances. Be careful not to run out of fuel and risk being captured.

RESPECT ALL AROUND
Although enemies, tank crews on each side had a grudging respect for their counterparts. Both sides had special uniforms to cope with the desert heat.

"D-d-d-did you just hear something out there?"

TENSION AND TREPIDATION
The anxious wait before a battle was a time of tension and intense concentration for a tank crew. The mental strain of battle was heightened by being in such an enclosed space.

19

Clash of armour

You and your crew in this Soviet tank are about to see your first fighting. You're all young – the gunner is only nineteen – but you're ready to take on the might of the German Army. Word has reached you that your home city, Kursk, in western Russia is a main target for those German tanks.

It's a little scary to hear some of the other soldiers talking about the blitzkrieg that raced across France three years ago in 1940, sweeping the French aside. Now Hitler wants to throw all his might into wiping out not just Kursk, but the whole Soviet defence. Yet again he's planning to plough through the enemy, this time to create a route east to the oil fields of Asia.

Luckily, there's a delay. Has Hitler changed his mind? People say he's a bit crazy and decides things on a whim. No matter, the delay means extra time for the Soviet army to build up its forces. 'Come and get us – we're not afraid', is the phrase on everyone's lips, but deep down you all know that the Battle of Kursk is going to be a fight to the death.

Battle of Kursk, 1943

SOVIET SELF-PROPELLED ARTILLERY
Soviet artillery, most notably the SU-76, could take out German tanks – if the Germans didn't blast first. Who would be fastest on the trigger?

Handy Hint
Be careful walking along a battlefield. There could be mines buried there – some powerful enough to stop a tank.

TAKING REST
German tank crews sometimes stopped to join infantry soldiers during lulls in the fighting. Breaks in between the fighting – sometimes lasting only a few minutes – gave tank crews a rare chance to stretch and stand.

THE SOVIETS WIN
Soviet soldiers posed for photos in front of captured German tanks. The relief at the end of the battle was enormous. They knew that they'd saved not only their own lives, but their country.

ALEKSANDRA SAMUSENKO
Liaison Officer Aleksandra Samusenko was a tank commander and deputy commander of a Soviet tank battalion. She inspired her crew to stand their ground when they faced three German tanks at the Battle of Kursk.

21

Funny business?

It's hard to forget the disastrous Dieppe raid of August 1942. Troops and vehicles, including tanks, were stranded on the beach while the German artillery and aircraft bombarded them. That raid was meant to be a practice run for a much bigger landing in France, so people became very concerned.

Major-General Sir Percy Hobart studied what went wrong and saw that tanks needed to be far more versatile. He commanded adapted tanks that could cope with landing on French beaches and then take on the enemy. Some of the designs seemed a bit odd, so the nickname for these tanks, Hobart's Funnies, stuck. Could the Funnies turn the next big landing into a success?

This Canadian tank crew, including wounded soldiers, were among the nearly 2,000 Allied troops captured in the disastrous Dieppe raid of 1942.

BUILDING BRIDGES
A Churchill AVRE tank could be adapted to lay a bridge across a gulley or small river. This type of tank gained a reputation for being a 'do-it-all' armoured vehicle, forming the basis of many of the 'Funnies'.

D-Day and beyond

You're taking part in history's largest invasion by sea. The date – 6 June 1944 – will be remembered as D-Day. A combined force of British, Americans and Canadians has crossed the English Channel and landed on beaches in Normandy, a region of north-west France. Nearly 7,000 ships and landing vessels transport troops, weapons and vehicles to five beaches. Awaiting them is Hitler's 'Atlantic Wall', a massive series of defences to repel any attack.

Tanks will play a big part in this offensive and the attack will be a chance to put some of Hobart's ideas into practice on the beaches and on the battlefields beyond. Many of them have floated into shore, buoyed up by 'flotation skirts' which can be removed once the tanks are on land.

If all goes well, the Allies will break through the coastal defences and drive the Germans back. Negotiating the countryside beyond, with its hills, marshes and hedgerows, will be a further deadly challenge – even to a powerful tank.

Ahoy there! The tank's as dry as a bone.

FEARSOME FIREPOWER
The huge firepower of German anti-tank artillery could strike terror in an advancing Allied tank crew. The shells from these cannons could pierce the thickest tank armour on the battlefield.

Handy Hint
A tank that's hit can easily catch fire because of the ammunition. Make sure you bail quickly!

A CAPTURED PRIZE
German Tiger tanks were abandoned as their crews fled before the Allied advance from the beaches through the Norman countryside.

A BREAK IN THE FIGHTING
Breaks in fighting gave tank crews the chance to swap stories with others while snatching a quick meal outside.

GOOD LUCK
Mascots brought tank crews luck. A grateful Dutch villager gave this teddy to a British tank crew in 1944. The bear travelled across Europe.

RACIAL PREJUDICE
The African-American crews of the US 761st Tank Battalion fought two enemies – Hitler's Germany and racial prejudice back home.

New battlefields

As the war progressed, tanks wound up in the thick of fighting – in open ground, arid deserts, snowy forests and even in city streets. Tanks had to crash through thick jungles in Asia and on Pacific islands. As soon as the crews, cooped up in steamy tanks, dared to open the hatch, they faced risks – from malarial mosquitoes, venomous snakes and enemy snipers in the undergrowth.

The courage and endurance of tank crews helped the Allies defeat Germany (in May 1945) and Japan (in August 1945). They helped win the war and restore the peace.

FLAMETHROWING TANKS
These were terrifying additions to the fighting forces in the Second World War. This American tank used flames to flush out the enemy on a Pacific island.

M4 SHERMAN
The crew of an American M4 Sherman tank proudly relieved their fellow US soldiers who'd been surrounded by the enemy in Bastogne, Belgium, in December 1944.

Handy Hint

Be prepared to get your hands dirty. Even the best tanks can get stuck in mud and need some coaxing to get free.

SOVIET TANKS TAKE BERLIN
Soviet tanks entered Berlin in April 1945 during the last days of the war. More than 30 million Soviet citizens had died in the war.

TANK VICTORY
A 1945 victory parade in Pennsylvania gratefully acknowledged the role of tanks and the extrodinary bravery of their crews.

Life in a tank

What is it like when the main gun is fired?

Most tank crew members nowadays have ear protection as a rule – because just running a tank is noisy. The sound of the main gun is heard more outside than inside the tank because the metal surrounding the cannon is so thick. Crew members do feel a shock wave, though. One American tank commander described the feeling of having his head outside the hatch when the main gun was fired: 'The shock from the main gun when it goes over you and expands out in that spherical pattern, rattles your teeth. It blows all of the hair on your face back. I wore a moustache, I always wore a moustache, so it would blow my moustache back and flatten down my eyebrows.'

Can you cook in a tank?

Open flames in an enclosed space, surrounded by powerful explosives, are never a good idea. Tank crews have traditionally made do with army rations, mainly tinned and dried food, and only been able to cook when they've left the vehicle. British crews had their wish come true in the 1950s. Their new Centurion tanks were fitted with boiling vessels (large kettles) that allowed them to cook and brew tea in safety.

Handy hint

You can stop crossing your legs. Some of today's futuristic tanks now have toilets!

What does it smell like inside a tank?

It's easy to imagine that even with deodorant, a crew squashed together in an enclosed metal box will begin to smell a little... powerful. Crews will take every opportunity to open the hatches but during combat – when nerves make people sweat even more – those hatches remain firmly shut, and the odour increases.

Glossary

AI Artificial Intelligence, a branch of computer science that develops programs to enable computers to make decisions independently.

Alliance A group of countries that agree to support each other.

American Civil War A war that lasted from 1861–1865 in which some southern states of the USA fought to form their own country but were defeated.

Ammunition Objects such as bullets or cannon shells that can be shot from a weapon.

Arid Extremely dry because of lack of rainfall.

Artillery Large weapons used in battle.

Assembly line A method of producing goods in a factory by moving them along a line of workers who specialise in one part of the assembly.

Battalion A large group of soldiers under a single command.

Blitzkrieg Literally 'lightning war' in German, a term to describe a sudden attack moving quickly across an invaded country.

Blueprint A detailed diagram of a product, usually with white lines against a blue background.

Cavalry The branch of the army whose soldiers are mounted on horses.

Civilian Someone who is not part of the armed forces.

Civil war A war between two groups within a single country.

GPS Short for Global Positioning System, a method of finding locations or navigating using signals from satellites orbiting the Earth.

Great War Another name for the First World War (1914–1918).

Hatch An opening in the side or on top of a vehicle such as a tank, to allow people or goods to pass through.

Hoisted When something is lifted up using a mechanical device.

Ironclad Covered in protective metal.

Kurdish Coming from Kurdistan, a region that extends into parts of Turkey, Syria, Iraq and Iran.

Mascot A person or thing that is supposed to bring good luck.

Negotiating Driving or riding safely through an area.

Periscope A tube containing two mirrors to allow a person to view around obstacles.

Projectile The part of a piece of ammunition that is shot from a weapon.

Racial prejudice The belief that some people are better – or worse – than others because of the colour of their skin.

Rallies Organised gatherings of people who show their support for a cause or a political leader.

Republican During the Spanish Civil War (1936–1939), this was someone who supported the elected government, known as the Republic, against those who wanted to overthrow it.

Revolution The forceful overthrow of a government, replacing it with a new government.

Riveter Someone who operates the machine that fastens (rivets) pieces of metal together.

Terrain The natural features of the land or ground in an area.

Trench warfare A type of fighting in which two armies face each other on a battlefield, using a series of long ditches (trenches) for protection against enemy shooting.

Venomous Able to injure or kill humans by injecting a harmful liquid, called venom.

Versatile Able to perform a wide variety of tasks.

Index

A
Afrika Korps 18, 35
AI 28, 37
Allies 1, 14, 16, 23–24, 26, 35
armour 5, 7–9, 11–13, 16, 18, 20, 22, 25, 29–30, 34, 36
Atlantic Wall 24

B
Battle Car 6, 34
Belgium 26, 36
blitzkrieg 13–14, 20, 31, 35
British Expeditionary Force 14
Burstyn, Gunther 8

C
Calais 23
cavalry 10
Central Powers 1
Centurion tanks 33
Churchill tanks 14, 22–23
commander 16–17, 21, 32

D
da Vinci, Leonardo 6, 34
D-Day 1, 24, 35
de Mole, Lancelot 9
Dieppe 1, 22
digital communications 31
driver 16–18
Dunkirk 1, 14, 35

E
El Alamein 18, 36
English Channel 14, 24

F
firepower 5–6, 8, 25
First World War 1, 8, 10, 12, 16, 30–31, 36
flotation skirts 24

G
GPS 28
gunner 16–17, 20

H
hatch 16, 26, 32–33
Hitler, Adolf 12, 20, 24–25
Hobart's Funnies 22
Hobart, Sir Percy 22, 24
Hornsby Tractor 8
hull machine gunner 17

I
Internet 31
Ironclads 7, 34
Iron Triangle 8

J
Japan 1, 12, 26, 35
Jones, Mavis 14

K
Kursk 20–21, 35–36

L
Little Willie 8–9
loader 16–17

M
mobility 5–6, 8
Montgomery, Sir Bernard 18

N
Normandy 24, 36

P
protection 5–8, 32

R
radios 13, 30–31
railway 15
recruits 29
Republicans 1
robot tanks 29
Rommel, Erwin 18, 35
Rosie the Riveter 15
Royal Tank Corps 12

S
Samusenko, Aleksandra 21
semaphores 30
Sherman tanks 16, 26
shock wave 32
Spanish Civil War 12

T
Tiger tank 25
toilet 11, 28–29
trench warfare 7–8, 10–11, 36
tsar 1

U
US 761st Tank Battalion 25

W
Wells, H. G. 7
Women 14–15, 28–29

My name is Ernie, and I'm a guide here at The Tank Museum. This job was made for me. You could say it's all in the family. You see, I'm named after my dad, who was part of a tank crew in the Second World War. I was a year old when the war started and didn't see much of Dad until it ended in 1945. And Dad's dad, my grandfather Bill, rode a tank in the First World War, when the first tanks saw action.

That war (1914–1918) involved two great alliances. On one side were the Allies, including Great Britain, Russia, France, Italy, Greece, the United States and several other countries. Opposing them were the Central Powers – Germany, Turkey, Austria-Hungary and Bulgaria. Of course people didn't call it the First World War at the time. They simply called it the Great War because it saw so many soldiers, from so many countries, fighting – sometimes to gain just a few metres of land.

Midway through the war, in 1917, Russia had a revolution. The country changed its name to the Soviet Union and executed its leader, the Tsar, and his family. The Allies finally defeated the Central Powers in 1918. The 1920s were relatively peaceful, but by the 1930s it became clear that Germany was building a powerful military. In the late 1930s, it sent military equipment to help Spanish military forces trying to overthrow the Republican government. The Soviet Union also sent aid – but they supported the Republicans in that Civil War.

Germany invaded Poland in 1939, triggering the Second World War. Again, it was part of an alliance – the Axis powers – along with Japan and Italy. Great Britain, Canada, the Soviet Union and the United States – known simply as the Allies – opposed them. Germany swept through northern Europe in 1940. British forces were pushed back from the continent in 1940, retreating to Britain from the French port of Dunkirk.

In August 1942 the Allies tried to land troops and equipment at Dieppe, along the French coast, but it was a disastrous failure. However, they learned many lessons from that defeat and launched a massive (and successful) landing – known as D-Day – in France in 1944. That signalled the beginning of the end of the war, and Allied success, but fighting had been taking place and continued in many other areas, such as the Soviet Union and Asia.

The following pages concentrate on that Second World War and the special role that tanks and their crews played throughout it. At the end of the book, pages 30–36, you'll see maps, a timeline and more text to help you understand what happened when – and where – in the Second World War, and how tanks and their crews were at the heart of it.

First published in the UK in 2011 by The Salariya Book Company Ltd
This edition published in the UK in 2024 by Hatch Press,
an imprint of Bonnier Books UK
4th Floor, Victoria House
Bloomsbury Square, London WC1B 4DA
Owned by Bonnier Books
Sveavägen 56, Stockholm, Sweden
www.bonnierbooks.co.uk

Copyright © 2024 by Hatch Press

1 3 5 7 9 10 8 6 4 2

All rights reserved

ISBN 978-1-80078-942-5

Printed in United Kingdom